MACRO-EVOLUTION

A Theory- or Evidence-Based Science

MACRO-EVOLUTION

A Theory- or Evidence-Based Science

TIM HEATLEY ACMA CGMA

Copyright © 2020 Tim Heatley
All rights reserved.

ISBN: 9798680236420
Imprint: Independently published

List of Contents

Page	Subject
1	Introduction
5	Irreducible Complexity
15	Mutations
19	Fossils
21	Genetic Entropy
23	What the Experts Say
28	Conclusion
30	Bibliography
32	About the Author

Introduction

I feel a little presumptuous about writing this booklet as I don't have any advanced science qualifications. However, I have come across a sufficient number of scientists who would wish for such a publication but who fear the personal repercussions if they were to write on this matter. This is not a book about theology, nor does it matter much from a religious point of view as many ministers believe in something called "theistic evolution", which basically means that there is an intelligent design of some sort which created a system that involves evolution, humans being the highest end of that process. However, I do feel it matters from an educational point of view and from the point of scientific integrity. The main thing that motivated me to write this booklet was when I attended a lecture by a professor of chemistry who, while speaking on the subject of evolution, showed diagrams of apes gradually transitioning into a human that resembled an African gentleman. I felt this was insensitive to our BAME colleagues but also that the talk lacked subtlety and adequate referencing. When I wrote to the professor to discuss this issue he was very defensive and unwilling to receive questions about his talk.

So let's investigate this whole question further. There are various types of evolutionary thinking and so here are some of them with their definitions:

Microevolution is the change in allele frequencies (*one of two or more alternative forms of a gene, occupying the same position locus on paired chromosomes and controlling the same inherited characteristic*) that occurs over time within a population. This change is due to four different processes: mutation, selection, gene flow and genetic drift. This change happens over a relatively short amount of time compared to the changes termed macroevolution. (*Wikipedia*)

Macroevolution is evolution theorized to occur over a long period of time, producing major changes in species and other taxonomic groups. (*Encarta English Dictionary UK*)

In short, then, microevolution is the sort that we see in the form of natural selection within a species and involves the survival of the fittest, but macroevolution involves a change from one species to another often thought to happen by mutation.

This is not to be confused with **metamorphosis** which is defined here as follows: a complete or marked change in the form of an animal as it develops into an adult, e.g. the change from tadpole to frog or from caterpillar to butterfly. (*Encarta English Dictionary UK*)

This booklet is not intended to answer the question of how life came into existence. That is more of a philosophical or theological question. Science itself is not in question here as that is purely objective and a basis for all investigation. It is the interpretation of science or the formulation of theories and hypotheses that matters and their method of presentation.

In some discussions the age of the universe is debated, as a young age would mean more  difficulty for the evolution camp. However, I will not deal with this here as I believe that it is very difficult to calculate this either from a scientific point of view or from a historical or biblical perspective.

I also believe that the measurement of time becomes problematic from the moment of big bang since the forces involved and the speeds of the expanding galaxies thereafter will to some extent distort time as we know it (see *Why does E=mc²* by Brian Cox and Jeff Forshaw).

Irreducible Complexity: The Challenge to the Darwinian Evolutionary Explanations of many Biochemical Structures

> "To suppose that the eye with all its inimitable contrivances for adjusting the focus to different distances, for admitting different amounts of light, and for the correction of spherical and chromatic aberration could have been formed by natural selection seems, I freely confess, absurd in the highest degree."

Charles Darwin in the *Origin of Species,* J M Dent and Sons Ltd, London, 1971 p.167 (page 18 of the revised quote book).

Also from the *Origin of Species:*

> "If it could be demonstrated that any complex organ existed which could not possibly have been formed by numerous, successive, slight modifications, my theory would absolutely break down."

With these statements, Charles Darwin provided a criterion by which his theory of evolution could be falsified. The logic was simple: since evolution is a gradual process in which slight modifications produce advantages for survival, it cannot produce complex structures in a short amount of time. It's a step-by-step process which may gradually build up and modify complex structures, but it cannot produce them suddenly.

Darwin, meet Michael Behe, biochemical researcher and professor at Lehigh University in Pennsylvania. Michael Behe claims to have shown exactly what Darwin claimed would destroy the theory of evolution, through a concept he calls "irreducible complexity." In simple terms, this idea applies to any system of interacting parts in which the removal of any one part destroys the function of the entire system. An irreducibly complex system then, requires each and every component to be in place before it will function.

As a simple example of irreducible complexity, Behe presents the humble mousetrap.

It contains five interdependent parts which allow it to catch mice: the wooden platform, the spring, the hammer (the bar which crushes the mouse against the wooden base), the holding bar, and a catch. Each of these components is absolutely essential for the function of the mousetrap. For instance, if you remove the catch, you cannot set the trap and it will never catch mice, no matter how long they may dance over the contraption. Remove the spring, and the hammer will flop uselessly back and forth - certainly not much of a threat to the little rodents. Of course, removal of the holding bar will ensure that the trap never catches anything because there will again be no way to arm the system.

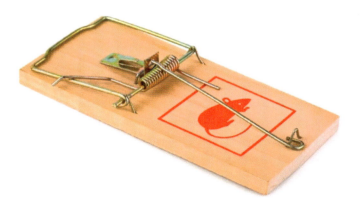

Now, note what this implies: an irreducibly complex system cannot come about in a gradual manner. One cannot begin with a wooden platform and catch a few mice, then add a spring, catching a few more mice than before, etc. No, all the components must be in place before it functions at all. A step-by-step approach to constructing such a system will result in a useless system until all the components have been added. The system requires all the components to be added at the same time, in the right configuration, before it works at all.

How does irreducible complexity apply to biology? Behe notes that early this century, before biologists really understood the cell, they had a very simplistic model of its inner workings. Without the electron microscopes and other advanced techniques that now allow scientists to peer into the inner workings of the cell, it was assumed that the cell was a fairly simple blob of protoplasm. The living cell was a "black box" - something that could be observed to perform various functions while its inner workings were unknown and mysterious. Therefore, it was easy, and justifiable, to assume that the cell was a simple collection of molecules. But not anymore. Technological advances have provided detailed information about the inner workings of the cell. Michael Denton, in his book *Evolution: A Theory in Crisis*, states

> "Although the tiniest bacterial cells are incredibly small, weighing less than 10^{-12} grams, each is in effect a veritable microminiaturized factory containing thousands of exquisitely designed pieces of intricate molecular machinery, made up altogether of one hundred thousand million atoms, far more complicated than any machine built by man and absolutely without parallel in the non-living world."

In a word, the cell is complicated. Very complicated.

In fact, Michael Behe asserts that the complicated biological structures in a cell exhibit the exact same irreducible complexity that we saw in the mousetrap example. In other words, they are all-or-nothing: either everything is there and it works, or something is missing and it doesn't work. As we saw before, such a system cannot be constructed in a gradual manner - it simply won't work until all the components are present, and Darwinism has no mechanism for adding all the components at once. Remember, Darwin's mechanism is one of gradual mutations leading to improved fitness and survival. A less-than-complete system of this nature simply will not function, and it certainly won't help the organism to survive. Indeed, having a half-formed and hence non-functional system would actually hinder survival and would be selected against. But Behe is not the only scientist to recognize irreducible complexity in nature. In 1986, Michael J Katz, in his *Templets and the explanation of complex patterns* (Cambridge University Press, 1986) writes:

> "In the natural world, there are many pattern-assembly systems for which there is no simple explanation. There are useful scientific explanations for these complex systems, but the final patterns that they produce are so heterogeneous that they cannot effectively be reduced to smaller or less intricate predecessor components. As I will argue ... these patterns are, in a fundamental sense, irreducibly complex..."

Katz continues that this sort of complexity is found in biology:

> "Cells and organisms are quite complex by all pattern criteria. They are built of heterogeneous elements arranged in heterogeneous configurations, and they do not self-assemble. One cannot stir together the parts of a cell or of an organism and spontaneously assemble a neuron or a walrus: to create a cell or an organism one needs a pre-existing cell or a pre-existing organism, with its attendant

complex templates. A fundamental characteristic of the biological realm is that organisms are complex patterns, and, for its creation, life requires extensive, and essentially maximal, templates."

Behe presents several examples of irreducibly complex systems to prove his point, but I'll just focus on one: the cilium. Cilia are hair-like structures, which are used by animals and plants to move fluid over various surfaces (for example, cilia in your respiratory tree sweep mucous towards the throat and thus promote elimination of contaminants) and by single-celled organisms to move through water. Cilia are like "oars" which contain their own mechanism for bending. That mechanism involves tiny rod-like structures called microtubules that are arranged in a ring. Adjacent microtubules are connected to each other by two types of "bridges" - a flexible linker bridge and an arm that can "walk" up the neighbouring microtubule. The cilia bends by activating the "walker" arms, and the sliding motion that this tends to generate is converted to a bending motion by the flexible linker bridges.

Thus, the cilium has several essential components: stiff microtubules, linker bridges, and the "motors" in the form of walker arms. While my description is greatly simplified (Behe notes that over 200 separate proteins have been identified in this particular system), these three components form the basic system, and show what is required for functionality. For without one of these components, the system simply will not function. We can't evolve a cilium by starting with microtubules alone, because the microtubules will be fixed and rigid - not much good for moving around. Adding the flexible linker bridges to the system will not do any good either; there is still no motor and the cilia still will not bend. If we have microtubules and the walker arms (the motors) but no flexible linker arms, the microtubules will keep on sliding past each other till they float away from each other and are lost.

This is only one of many biochemical systems that Behe discusses in his book, *Darwin's Black Box*. Other examples of irreducible complexity include the light-sensing system in animal eyes, the transport system within the cell, the bacterial flagellum, and the blood clotting system. All consist of a very complex system of interacting parts which cannot be simplified while maintaining functionality.

Since the publication of *Darwin's Black Box*, Behe has refined the definition of irreducible complexity. In 1996 he wrote that "any precursor to an irreducibly complex system that is missing a part is by definition non-functional." (Behe, M, 1996b. Evidence for Intelligent Design from Biochemistry, a speech given at the Discovery Institute's God & Culture Conference, 10th August 1996, Seattle, WA. http://www.arn.org/docs/behe/mb_idfrombiochemistry.htm).

By defining irreducible complexity in terms of "non-functionality," Behe casts light on the fundamental problem with evolutionary theory: evolution cannot produce something where there would be a non-functional intermediate. Natural selection only preserves or "selects" those structures which are functional. If it is not functional, it cannot be naturally selected. Thus, Behe's latest definition of irreducible complexity is as follows:

> "An irreducibly complex evolutionary pathway is one that contains one or more unselected steps (that is, one or more necessary-but-unselected mutations). The degree of irreducible complexity is the number of unselected steps in the pathway." (A Response to Critics of Darwin's Black Box, by Michael Behe, PCID, Volume 1.1, January February March 2002; iscid.org/)

Evolution simply cannot produce complex structures in a single generation as would be required for the formation of irreducibly complex systems. To imagine that a chance set of mutations would produce all 200 proteins required for cilia function in a single generation stretches the imagination beyond the breaking point. And yet, producing one or a few of these proteins at a time, in standard Darwinian fashion, would convey no survival advantage because those few proteins would have no function - indeed, they would constitute a waste of energy for the cell to even produce. Darwin recognized this as a potent threat to his theory of evolution - the issue that could completely disprove his idea. So the question must be raised: Has Darwin's theory of evolution "absolutely broken down?" According to Professor Michael Behe, the answer is a resounding "Yes."

Mutations

Adapted from *Creation and Change* by Douglas F Kelly 1997. Douglas Kelly is a Professor of Theology who has studied science as well as philosophy:

Gene mutations are said to occur when individual genes are damaged from exposure to heat, chemicals, or radiation. Chromosome mutations occur when sections of the DNA are duplicated, inverted, lost, or moved to another place in the DNA molecule.

Thus the central mechanism of evolution is mutations. Therefore, a great deal of work has been done in the last fifty years to see if they establish evolutionary theory. The fruit fly in particular has been studied intensively with this idea in mind. It has been the subject of many experiments because its short life-span allows scientists to observe many generations. In addition, the flies have been bombarded with radiation to increase the rate of mutations. Scientists now have a pretty clear idea of what kind of mutations can occur.

Mutations do not create new structures. They merely alter existing ones. Mutations have produced, for example, crumpled, oversized and undersized wings. They have produced double sets of wings, but they have not created a new wing. Nor have they transformed the fruit fly into a new kind of insect. Experiments have simply produced variations within the fruit fly species.

Mutations are quite rare. This is fortunate, for in virtually all instances they are harmful. Recall that the DNA is a molecular message. A mutation is a random change in the message, akin to a typing error. Typing errors rarely improve the quality of a written message: if too many occur, they may even destroy the information contained in it. Likewise, mutations rarely improve the quality of DNA message, and too many may even be lethal.

How likely is it that random mutations will come together and coordinate to form just one new structure? Let's say the formations of an insect wing require only five genes (a very low estimate). Most mutations are harmful and scientists estimate that only one in 1,000 is not. The probability of two non-harmful mutations occurring is one in one thousand million. For all practical purposes, there is no chance that all five mutations will occur within the life cycle of a single organism. So far, we have discussed the possibility of the random formation of only one structure. Yet, an organism is made of many structures that must appear at the same time and working together in an integrated whole.

Thus genetic facts militate against evolution being possible through either recombination or mutations. The concept of the upward evolutionary scale of life is not grounded in empirical science; it is actually contrary to it. This means that the theory of evolution is really philosophical, not operational science. In the words of the French biologist, Remy Chauvin, Professor in the Laboratory of Animal Sociology at Rene Descartes University in Paris: "I say, and underline, the fact that if mass preconceived ideas did not exist, everyone would admit that since these forms of animal life which mutate very rapidly have remained the same during tens of millions of generations, mutation could not be considered the motor of evolution. This is a matter of good sense, but given the strength of prejudice within science as everywhere else, good sense loses its case in court".

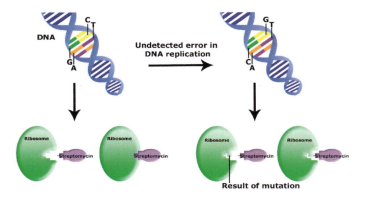

Lynn Margullis, Professor of Biology at the University of Massachusetts, has stated the position even more bluntly. She shows that molecular biology has yet been unable to demonstrate the formation of a single new species by mutations, and concludes that neo-Darwinism is a "minor twentieth-century religious sect within the sprawling religious persuasion of Anglo-Saxon biology".

Fossils

It has been said of Darwin that he referenced his theories by saying said that if these were to be proven then the geological layers would be awash with fossils of transitional forms between species. It's now 150 years since the *Origin of Species* was published and so far not one such legitimate fossil has been found despite the discovery of trillions of fossils around the earth and thousands of these now

dug up and analysed. There have been a few hoaxes, but that's all they were, now discredited. This much sort-after missing link which would demonstrate a transition from species to species is still just that, the missing link.

Genetic Entropy

Entropy: the second law of thermodynamics says that entropy always increases with time. The dictionary definition (Google Dictionary) of entropy is a lack of order or predictability or a gradual decline into disorder.

Genetic Entropy Confirmed Science 3 June 2011: Vol. 332 no. 6034 pp. 1193-1196 DOI: 10.1126/science.1203801; *Negative Epistasis Between Beneficial Mutations in an Evolving Bacterial Population,* Aisha I Khan. This publication evidences a gradual wearing down of the genes, producing weaker generations each cycle. Perhaps this accounts for why people appeared to be tougher and even live longer in some reports from ages passed. So in a way you could call this process "devolution".

N.B. Epistasis is the nonappearance of a characteristic determined by one gene because it has been suppressed or masked by the activity of another gene. (Encarta Dictionary)

What the Experts Say

It's interesting that Phillip E Johnson, Professor of Law at the University of California at Berkely, whose speciality is the evaluation of evidences, has written a monograph demonstrating the faith-basis of evolution: *Evolution as Dogma: the Establishment of Naturalism.* He says:

> "What the science educators propose to teach us as 'evolution', and label as fact, is based not upon any incontrovertible empirical evidence, but upon a highly controversial philosophical pre-supposition".

It is also noteworthy that some highly-qualified scientists have challenged Darwin's theories as implausible. We have a number in the UK and notable examples of these are: a professor of thermodynamics in Leicestershire, a professor of combustion theory in Yorks, and a reader in engineering design in Avon. I have excluded the names here to protect these people from the persecution that often follows those that challenge current scientific theory.

Then there was the late great Professor Wilder-Smith who had a Ph.D. in physical organic chemistry at University of Reading, England (1941) and Dr.es.Sc. in pharmacological sciences from Eidgenossische Technische Hochschule (Swiss Federal Institute of Technology) in Zurich, D.Sc. in pharmacological sciences from University of Geneva (1964), F.R.I.C. (Fellow of the Royal Institute of Chemistry) and Professorships held at numerous institutions including: University of Illinois Medical School Center (Visiting Full Professor of Pharmacology, 1959-61, received 3 "Golden Apple" awards for the best course of lectures), University of Geneva School of Medicine, University of Bergen (Norway) School of Medicine, Hacettepe University (Ankara, Turkey) Medical School, etc. He was former Director of Research for a Swiss pharmaceutical company. He presented the 1986 Huxley Memorial Lecture at the invitation of the University of Oxford and was author or co-author of over 70 scientific publications and more than 30 books, published in 17 languages.

I have some scientist/academic friends who are also of this persuasion. Indeed, a local scientist acquaintance of mine who is a doctor of parasitology has some interesting points to make. When I talked to him about the recent work on genetic entropy he made the following observation:

> "Have we ever seriously thought about where DNA and proteins came from and how they got organised in the way they have to enable the whole machinery/system of protein synthesis and DNA replication in which context mutations take place? This whole system of protein synthesis and DNA replication involves hundreds of enzymes ... which came first, the proteins or the DNA? If DNA, then how did it replicate and assemble without the whole protein framework upon which that depends? If proteins came first, how did they get synthesized without the coded DNA/RNA templates upon which protein synthesis depends? These are fundamentals of biology even before we get to mutations and structures and genes and alleles".

In this context refer to the last article on genetic entropy which demonstrates that organisms are winding down (devolving) and not evolving.

Finally, Sir Fred Hoyle:

> "If you stir up simple nonorganic molecules like water, ammonia, methane, carbon dioxide and hydrogen cyanide with almost any form of intense energy ... some of the molecules reassemble themselves into amino acids ... demonstrated ... by Stanley Miller and Harold Urey. The ... building blocks of proteins can therefore be produced by natural means. But this is far from proving that life could have evolved in this way. *No one has shown that the correct arrangements of amino acids*, like the orderings in enzymes, can be produced by this method. .... A junkyard *contains all the bits and pieces of a Boeing 747*, dismembered and in disarray. *A whirlwind* happens to blow through the yard. *What is the chance* that after its passage *a fully assembled 747*, ready to fly, will be found standing there? *So small as to be negligible*, even if a tornado were to blow through enough junkyards to fill the whole Universe."

(Hoyle, F, *The Intelligent Universe*, Michael Joseph: London, 1983, pp.18-19). Sir Fred

Hoyle FRS (24 June 1915 – 20 August 2001) was an English astronomer.

Conclusion

So I think it is a case of not letting the facts, or lack of them, spoil a good story. It is forbidden to officially discuss these things in class at school or university. A bit like the Soviet cover-up of the Chernobyl disaster, upsetting the status quo or criticising the state is not acceptable. Truth, although we live in a democracy, plays second place to expediency. When David Attenborough, one of the country's leading evolutionists, was presented with some of these facts on a BBC documentary he said that he could not accept them as this would mean having a different belief which was inconceivable to his world view. Good science is something that you are allowed to question and debate, particularly if this is done by qualified scientists. Yet, if a teacher or lecturer, however highly qualified, were to bring up any of these issues in front of a class then it could mean instant dismissal.

This is very strange and something that you would associate more with a communist regime than a free state. So all I am asking is for the UK to have the intellectual freedom to debate these issues in whatever context the citizens wish. By questioning and challenging is how we expand our understanding and knowledge and something we are encouraged to do in most other contexts. I am reminded of the song, "The King is in his altogether." The lyrics of that song would be a suitable conclusion to this booklet.

Bibliography:

Prof E H Andrews, *From Nothing to Nature,* Evangelical Press, 1978

Prof Michael Behe, *Darwin's Black Box: the Biomedical Challenge to Evolution,* The Free Press, New York 1996

Brian Cox and Jeff Forshaw, *Why Does E=mc²?* Da Capo Press 2009

Charles Darwin, *Origin of Species,* J M Dent and Sons Ltd, London, 1971

Fred Hoyle, *The Intelligent Universe,* Michael Joseph: London, 1983, pp.18-19)

Michael J Katz, *Templets and the explanation of complex patterns,* Cambridge University Press, 1986

Douglas F Kelly, *Creation and Change,* Christian Focus Publications, 1997

Aisha I Khan, Science 3, June 2011, Vol 332 no 6034, pp.1193-1196 DOI: 10.1126/science 1203801; *Negative Epistasis Between Beneficial Mutations in an Evolving Bacterial Population*

Stephen Law, *The Philosophy Gym,* Headline Book Publishing, 2003

About the Author:

Tim Heatley was born in 1949 in the city of Leicester. Now retired, he qualified as an accountant in 1974 and was motivated by his training and experience to investigate the integrity of accounts, theories or hypotheses that are not subject to proper scrutiny. Also, ever since his school days he has had a strong interest in science and science-related matters. Hence this booklet. He is married to Judith and they both now live in a small village in the county of Leicestershire. He has had articles published on theological issues although this small booklet is merely researching the integrity of certain scientific beliefs.

Printed in Dunstable, United Kingdom